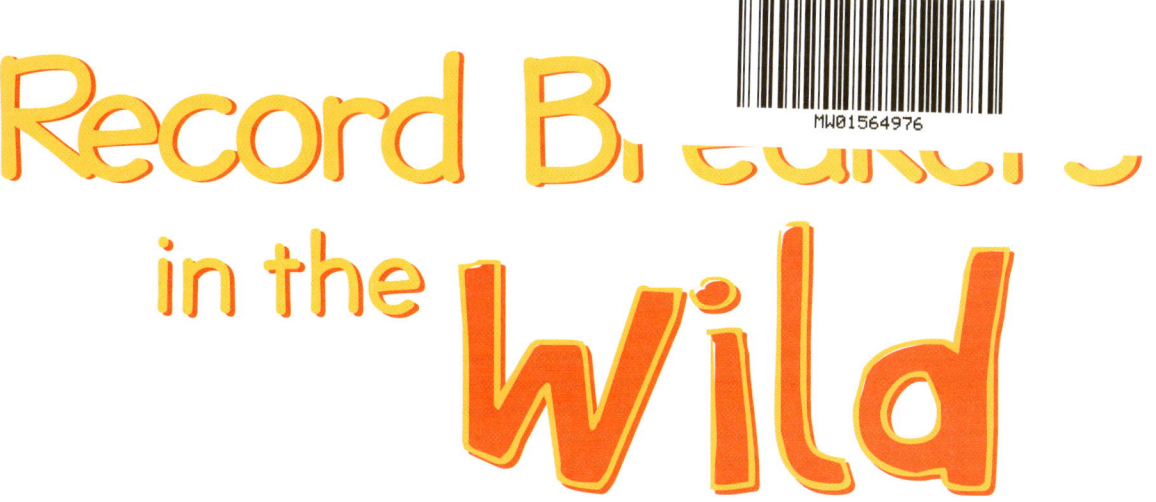

Record Breakers in the Wild

by Jordan Mack

SCHOLASTIC INC.

Photos ©: 3: Nigel Treblina/AFP/Getty Images; 4 main: Christian Darkin/Science Source; 5 top: FLPA/Alamy Images; 5 bottom: Don Johnston/age fotostock; 6 top: Dr. Merlin D. Tuttle/Science Source; 6 bottom: Lee Dalton/Alamy Images; 7 main: brittak/iStockphoto; 7 inset: Westend61/Getty Images; 9 top: NaluPhoto/iStockphoto; 9 bottom: Scubazoo/Alamy Images; 10 top: Anup Shah/Divine Images/Media Bakery; 10 center: AWEvans/iStock; 11 top: JodiJacobson/iStockphoto; 11 bottom center: Medioimages/Photoplay/Media Bakery; 12 bottom: Massimo De Medici/Getty Images; 13 top: nilky/iStockphoto; 16 blue whale: Christian Darkin/Science Source; 16 bumblebee bat: Dr. Merlin D. Tuttle/Science Source; 16 cheetah: Anup Shah/Divine Images/Media Bakery; 16 hummingbird: Lee Dalton/Alamy Images; 16 giraffe: brittak/iStockphoto; 16 sailfish: NaluPhoto/iStockphoto; 16 sea turtle: Scubazoo/Alamy Images; 16 sloth: JodiJacobson/iStockphoto; 16 zooplankton: FLPA/Alamy Images. All other images © Shutterstock.

No part of this publication may be reproduced in whole or in part, or stored in a retrieval system, or transmitted in any form or by any means, electronic, mechanical, photocopying, recording, or otherwise, without written permission of the publisher. For information regarding permission, write to Scholastic Inc., 557 Broadway, New York, NY 10012.

Portions previously published in Wild and Weird Record Breakers, copyright © 2012 by Clean Slate Press Ltd.

Copyright © 2020 by Scholastic Inc.
All rights reserved. Published by Scholastic Inc.
Printed in the U.S.A.

ISBN-13: 978-1-338-61663-7
ISBN-10: 1-338-61663-3

SCHOLASTIC and associated logos are trademarks and/or registered trademarks of Scholastic Inc.

5 6 7 8 9 10   132   29 28 27 26 25 24 23 22 21

Scholastic Inc., 557 Broadway, New York, NY 10012

Table of Contents

Measuring Up...................... **3**

The Mighty
and the Meek...................... **4**

Speedsters and
Slowpokes........................... **8**

Mega Muscles...................... **12**

Deadly Defenses................. **14**

Measuring Up

Wild animals come in all shapes and sizes, and with different sets of skills. Finding out which animal is the fastest, biggest, or slowest is not an easy task. Scientists work hard to discover which animals are record breakers. Take a look!

These scientists have a clever way to measure the length of this sea lion.

The Mighty and the Meek

The blue whale is the largest animal in the world. It can grow as long as 100 feet. That's about the distance between the bases on a baseball field. Weighing in at up to 200 tons, the blue whale is also the world's heaviest animal.

What is the largest animal

Heaviest Animal

Size Comparison

blue whale — elephant

Smallest Animals

Zooplankton are tiny animals, such as worms, mollusks, and shrimps.

The smallest ocean animals are zooplankton. Amazingly, the largest animal in the world feeds on one of the smallest animals. Blue whales eat tons of zooplankton daily.

Talk About Small!

This frog is a record breaker, too. It's one of the tiniest animals and the smallest frog in the world. It's no bigger than the size of a housefly!

Smallest Mammal

bumblebee bat

The bumblebee bat is only about one inch (2.5 cm) long.

The smallest bird in the world is the bee hummingbird. It weighs about as much as a penny! It flaps its wings around eighty times a second!

A bee hummingbird's egg is only the size of of a pea!

Smallest Bird

bee hummingbird

A giraffe is so tall, it can reach the tops of trees. The tallest giraffe ever measured was 19 feet, about half as tall as a telephone pole. Everything about the giraffe is long or tall, or both! Its neck is 6 feet, as tall as an adult man. And a giraffe's tongue is 20 inches long. That's about 5 times as long as an adult human tongue!

Tallest Land Animal

Giraffes use their record-breaking height and extra long, blue-black tongue to grab leaves at the tops of trees.

Speedsters and Slowpokes

The peregrine falcon is the fastest animal on Earth. This record breaker may reach a speed of 200 miles per hour (more than 320 kilometers per hour) as it swoops down on its prey from the sky. That's faster than the fastest roller coaster in the world!

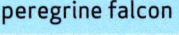
peregrine falcon

dragonfly

Dragonflies are the fastest flying insects in the world. The largest kind of dragonfly can fly up to 40 miles per hour (60 kilometers per hour).

Dorsal fin

sailfish

Fastest Swimmer

Sailfish are the fastest swimmers in the ocean. They get their name from their huge dorsal fin that looks just like a sail.

The leatherback turtle is a super-fast swimmer. It's the fastest reptile in the world.

Fastest Reptile

leatherback turtle

A cheetah can't run fast for a long time, but it can reach speeds of 70 miles per hour within three seconds. That's over the speed limit for driving a car on many highways. When running, cheetahs use their long tail to help steer and keep their balance.

Fastest Animal on Land

cheetah

Long-Distance Speedsters

Horses, pronghorn antelopes, and lions may not top a cheetah's 70-mile-an-hour speed, but they are amazing long-distance runners!

lion

pronghorn antelope

horse

Mega Muscles

An insect may be small, but that doesn't mean it can't be strong. The leafcutter ant is so strong, it can lift 50 times its own weight. And the strongest animal in the world is the rhinoceros beetle. It can carry 850 times its own weight. That means a beetle that weighs three ounces can carry about 160 pounds!

leafcutter ant

Strongest Animal

rhinoceros beetle

Slowest Animal

sloth

The sloth is the slowest animal on Earth. It moves so slowly that small green plants called algae grow on its coat. Sometimes moths live in the sloth's coat and feed on the algae.

Slow Going

While the sloth is the record breaker for the slowest animal, these four other animals are also known for how slowly they go.

snail

giant tortoise

sea star

Strongest Bite

saltwater crocodile

The saltwater crocodile has the strongest bite. It can break bones. The force of this croc's bite comes close to the mighty bite of the Tyrannosaurus rex!

Tyrannosaurus rex

Deadly Defenses

The box jellyfish is perhaps the most venomous animal in the world. More than 5,500 people have died from box jellyfish stings in the last 150 years.

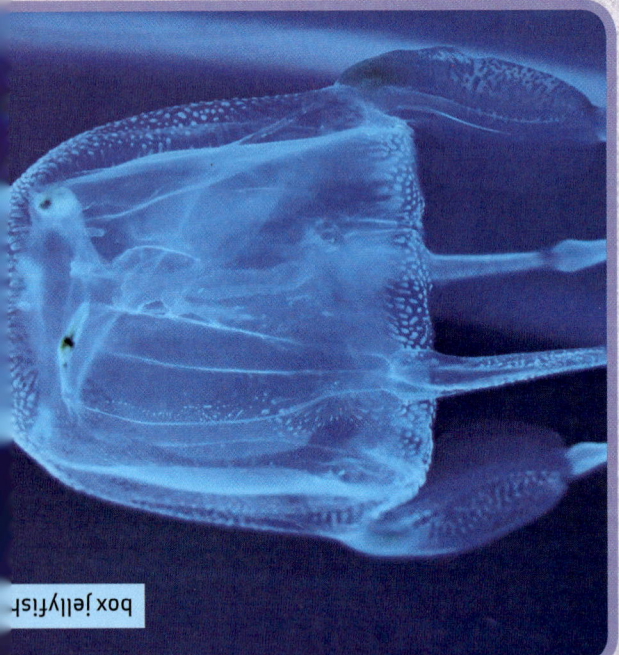

box jellyfish

Pretty Colors But Deadly!

There are over 175 kinds of poison dart frogs. Their skin can be bright orange, blue, red, or yellow. But whatever the color, a poison dart frog's skin oozes a deadly venom.

poison dart frog

The taipan is the world's most venomous snake. If you were bitten by a taipan, you would likely be dead within 45 minutes. However, the taipan is not aggressive. It avoids people. No deaths from taipan bites have been reported.

There are many amazing record breakers in the animal kingdom. Math and measuring help us see just how amazing they are.

Most Venomous Snake

taipan

Wild Animal Record Breakers!

Heaviest Animal:
blue whale
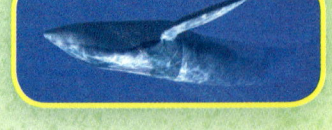

Fastest Animal:
peregrine falcon

Slowest Animal:
sloth

Smallest Animals:
zooplankton

Fastest Animal on Land:
cheetah

Strongest Animal:
rhinoceros beetle

Smallest Bird:
hummingbird

Fastest Flying Insect:
dragonfly

Strongest Bite:
saltwater crocodil

Smallest Mammal:
bumblebee bat

Fastest Swimmer:
sailfish

Most Venomous Animal:
box jellyfish

Tallest Land Animal:
giraffe

Fastest Reptile:
leatherback turtle

Most Venomous Snake:
taipan